100
things you should know about

MONKEYS & APES

100

things you should know about

MONKEYS & APES

Camilla de la Bedoyere

Consultant: Steve Parker

MC
PUBLISHERS

This 2009 edition published and distributed by:
Mason Crest Publishers Inc.
370 Reed Road, Broomall, Pennsylvania 19008
(866) MCP-BOOK (toll free)
www.masoncrest.com

Library of Congress Cataloging-in-Publication data is available
100 Things You Should Know About Monkeys and Apes
ISBN 978-1-4222-1522-7

100 Things You Should Know About - 10 Title Series
ISBN 978-1-4222-1517-3

Printed in the United States of America

First published in 2008 by Miles Kelly Publishing Ltd
Bardfield Centre, Great Bardfield, Essex, CM7 4SL

Copyright © Miles Kelly Publishing Ltd 2008

ACKNOWLEDGEMENTS

The publishers would like to thank the following artists
who have contributed to this book:
Mike Foster, Ian Jackson, Mike Saunders, Kim Thompson
Cover artwork: Ian Jackson
All other artworks are from the Miles Kelly Artwork Bank

The publishers would like to thank the following sources
for the use of their photographs:
Pages 6–7 Jurgen & Christine Sohns/FLPA; 8 Mark Newman/FLPA; 12 Jurgen & Christine Sohns/FLPA;
14–15 Jurgen & Christine Sohns/FLPA; 17 Photolibrary Group Ltd; 18 Pete Oford/naturepl.com;
19 PETE OXFORD/Minden Pictures/FLPA; 23 M. Watson/ardea.com; 25 Gerard Lacz/FLPA;
28 PETE OXFORD/Minden Pictures/FLPA; 29 (t) CYRIL RUOSO/JH EDITORIAL/Minden Pictures/FLPA;
29 (b) Simon Hosking/FLPA; 30 EcoView/Fotolia.com; 35 Photolibrary Group Ltd;
36 CYRIL RUOSO/JH EDITORIAL/Minden Pictures/FLPA; 38–39 Photolibrary Group Ltd;
42 Photolibrary Group Ltd; 44 (c) Getty Images; 46 NHPA/STEVE ROBINSON;
47 Frans Lanting/FLPA

All other photographs are from:
Castrol, Corel, digitalSTOCK, digitalvision, John Foxx, PhotoAlto,
PhotoDisc, PhotoEssentials, PhotoPro, Stockbyte

Contents

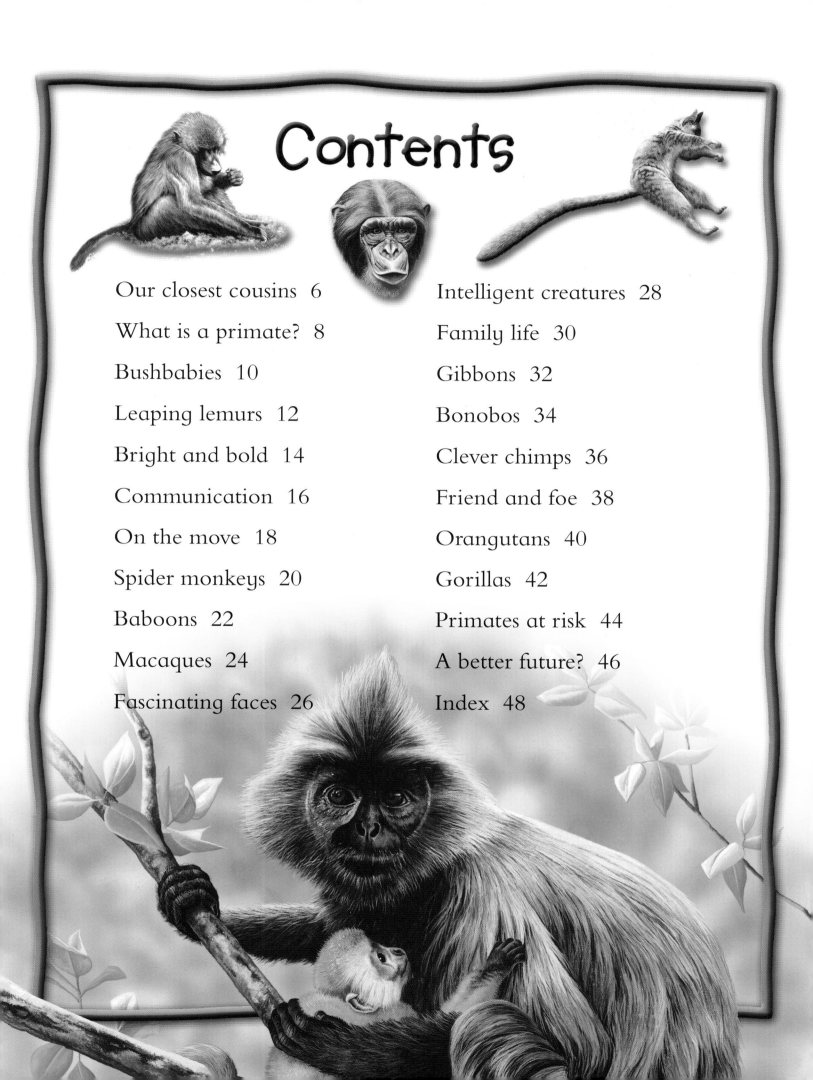

Our closest cousins

1 **Monkeys and apes are fascinating creatures.** They are our closest relatives and belong to a family of animals called primates. These animals can range in size from the tiny mouse lemur, which is no bigger than your hand, to the giant silverback gorilla. They may not all look alike, but primates share an important characteristic—they are very intelligent creatures.

These douc langur monkeys belong to the same family of animals as humans—primates.

What is a primate?

2 Primates such as monkeys and apes are covered in fur or hair and the young feed on their mothers' milk. They have big brains and can work out solutions to difficult problems. Primates have been known to learn new skills and teach them to their young.

3 Primates are mammals. This means that they have back bones and warm-blooded bodies. They are divided into three groups—prosimians, such as bushbabies, monkeys, such as baboons, and apes, such as gorillas.

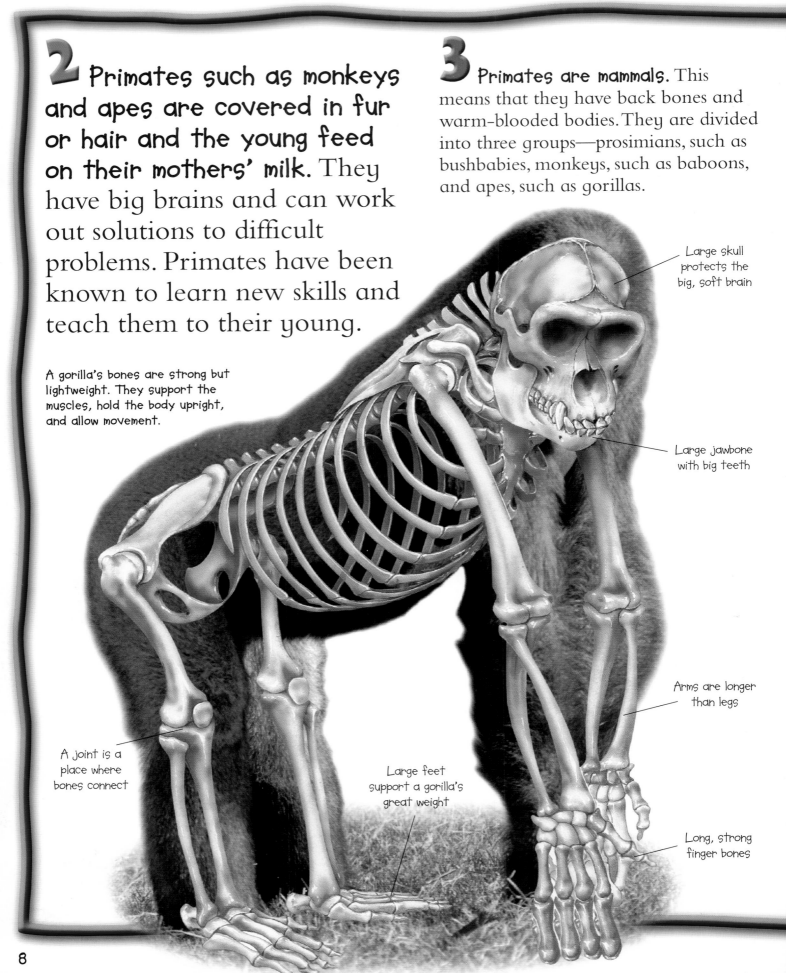

A gorilla's bones are strong but lightweight. They support the muscles, hold the body upright, and allow movement.

Large skull protects the big, soft brain

Large jawbone with big teeth

Arms are longer than legs

A joint is a place where bones connect

Large feet support a gorilla's great weight

Long, strong finger bones

4 Unlike many other animals, primates have large eyes at the front of their heads. This allows them to focus clearly on objects in front of them. Since most primates live in trees and leap between branches, this is a very useful feature. Unlike most other creatures, primates can see in color.

Primates' hands and feet can grab, hold, pinch, and probe. Most primates can grip objects and tools precisely in their hands.

Tarsier hand

Tarsier foot

Spider monkey hand

Spider monkey foot

Chimpanzee hand

Chimpanzee foot

6 Primates have hands that are very similar to ours. Instead of paws and claws, they have fingers and flat fingernails. They can bring their forefingers and thumbs together in a delicate pinching movement.

5 Primates prefer to live in groups. They often live with their families, or large groups of related families. Primates communicate with one another in lots of ways—using sound, scent, touch, and movement. Young primates usually stay with their families for years while they learn how to survive.

In many primate groups, such as the baboons shown here, adults help the mother by finding food and helping to look after the baby. Males will also gather food and play with the young.

Bushbabies

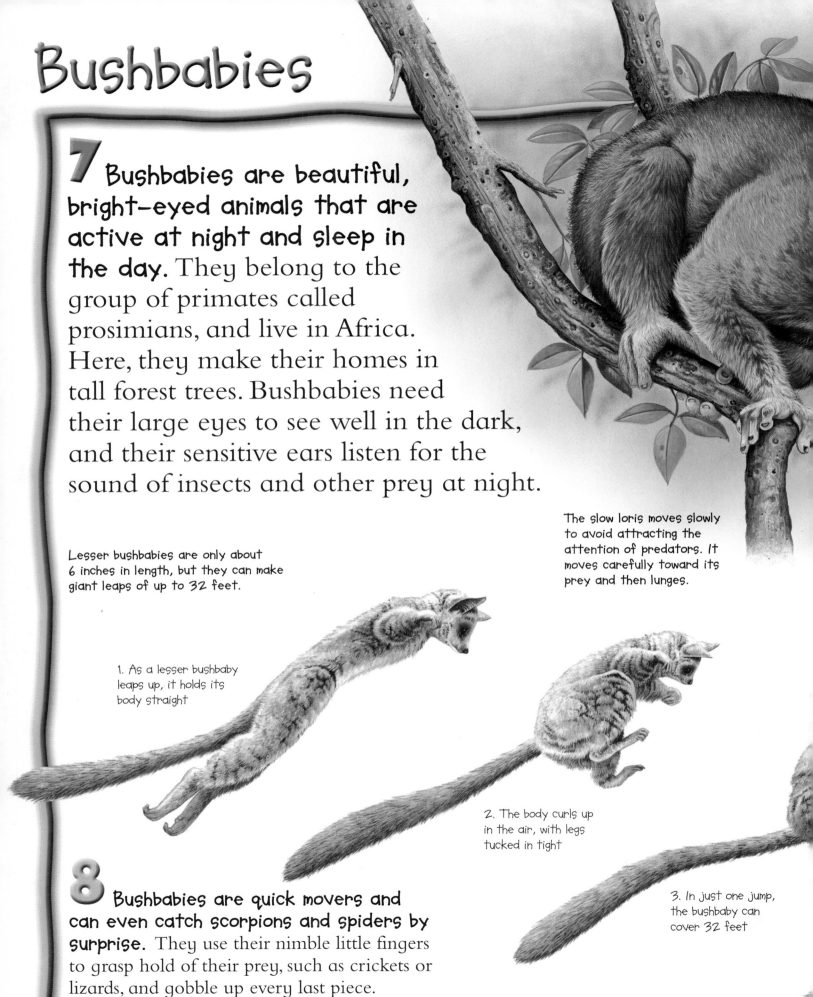

7 **Bushbabies are beautiful, bright-eyed animals that are active at night and sleep in the day.** They belong to the group of primates called prosimians, and live in Africa. Here, they make their homes in tall forest trees. Bushbabies need their large eyes to see well in the dark, and their sensitive ears listen for the sound of insects and other prey at night.

Lesser bushbabies are only about 6 inches in length, but they can make giant leaps of up to 32 feet.

The slow loris moves slowly to avoid attracting the attention of predators. It moves carefully toward its prey and then lunges.

1. As a lesser bushbaby leaps up, it holds its body straight

2. The body curls up in the air, with legs tucked in tight

3. In just one jump, the bushbaby can cover 32 feet

8 **Bushbabies are quick movers and can even catch scorpions and spiders by surprise.** They use their nimble little fingers to grasp hold of their prey, such as crickets or lizards, and gobble up every last piece.

Eyesight is very important to a tarsier and each eye is heavier than its brain. Tarsiers also have large ears that pick up the quietest of sounds.

9 **Lorises live in the forests of Southeast Asia.** Like bushbabies, they are closely related to monkeys and apes and lead similar lives. These furry animals spend most of their time in trees, where they move carefully and slowly through the leaves, searching for food. During the day they curl up in their hiding places and sleep.

10 **Tarsiers are odd-looking primates.** Their eyes are so big, they actually take up half the space in their heads! These primates can swivel their heads almost the whole way round, so they can watch what's going on behind them and look out for predators, or other animals to eat. They live in Southeast Asia and at night they leap with great ease through the trees.

4. Unlike most jumping animals, the bushbaby lands with its hind feet first

I DON'T BELIEVE IT!

Bushbabies have two tongues! When eating gum—a sticky substance made by trees—they use their teeth to scrape it off the bark. Then they wipe the gum off their teeth with their second tongue!

Leaping lemurs

11 Lemurs are long-legged primates that live in just one place on Earth—Madagascar. This large island in the Indian Ocean is home to lots of animals that aren't found anywhere else. Many of them, including lemurs, are dying out —partly because their forest homes are being cut down.

As well as being agile tree climbers, ring-tailed lemurs can move swiftly on the ground.

12 Ring-tailed lemurs are elegant, curious creatures. Unlike most other lemurs, they spend a lot of time on the ground and as they walk they hold their boldly-patterned tails high in the air. These primates rub their tails with smelly substances from under their arms. When two rival lemurs meet they wave their stinky tails at one another!

QUIZ

Most lemurs live in trees. What word is used to describe the place where an animal lives?

1. Halibut 2. Habit
3. Habitat

Answer:
3. Habitat

An aye-aye uses its long middle finger to probe into cracks in a tree and pick out tasty grubs to eat.

13 Newborn lemurs are soon strong enough to grasp onto their mothers' fur. As a female travels around between trees, her infant holds on tight, safe from predators. However the youngster is always at risk of falling to its death should it let go for a second.

14 The aye-aye is probably the ugliest primate. This shaggy-haired lemur builds its nest in the forks of trees and emerges at night to eat insects and fruit. Using its large ears, the aye-aye can hear beetles as they scratch around on the forest floor. An aye-aye's middle finger is unusually long—ideal for digging into wood and pulling out grubs to eat.

Sifakas can stand upright and run on their hind feet. They can leap 32 feet between branches.

15 Indris and sifakas are types of large lemur that are very close to extinction. They are gentle, plant-eating animals with loud voices, and can be heard calling from several miles away. Farmers and loggers are cutting down the forests where they live, and they are also hunted for meat.

Bright and bold

16 The beautiful golden lion tamarin is a familiar sight in zoos and wildlife parks. These brightly colored monkeys live in the tropical forests of Central and South America, but many of them are kept in captivity. At least 500 golden lion tamarins that were born in zoos have been released into the wild.

17 Tamarins belong to the group of primates called monkeys. All monkeys have tails, even if it's only a little stump. They are more intelligent than prosimians, but less intelligent than apes. Monkeys are divided into two groups—New World monkeys from the Americas and Old World monkeys from Europe, Asia, and Africa.

18 **The black-headed lion tamarin lives in just three places in Brazil.** It is one of the world's rarest monkeys and it's thought there are no more than 250 of them left alive, and even those few are fighting for survival. The reason for this is that their forest homes are being cut down so that wood from the trees can be sold.

Golden lion tamarins live in family groups of three to seven monkeys that stay together in one area, or territory. They fight with other groups that come into their territory.

19 **It's not just moms who look after babies—dads do too.** In most animal families, it's the mother that cares for the youngsters, but in tamarin families, the fathers share this important job. The mothers may feed the babies, but it's the fathers who carry the young on their backs, protecting them from owls, hawks, wild cats, and snakes.

I DON'T BELIEVE IT!

Not many monkeys have mustaches—but little emperor tamarins do! Other tamarins have crowns of white hair, beards, or tufts of fur on their ears.

Communication

20 Monkeys and apes may not use words to communicate, but they are good at letting each other know how they are feeling. Like humans, they can move their faces and make different noises to show their emotions.

CHIMP FACES

Do you look like a chimp? Stand in front of a mirror and copy the chimp faces shown on this page. Then make your own, "happy," "worried," and "pay me attention" faces. Is it easy to copy a chimp? Are your human expressions anything like the chimp's?

Play face

The chimp's eyes are relaxed and its mouth is open, but the top teeth are covered by the lip. Happy chimps can even make laughing noises.

Worried face

The lips are pulled right back and all the teeth are on show. The chimp makes high screeching sounds.

21 Leaving a strong smell behind you is a good way to let other animals know you've been around! Lots of creatures, not just monkeys and apes, leave their smell on trees and the ground to warn other animals to stay away. Monkey smells are made in special body parts called glands, which are often around the animals' rumps (bottoms).

If a young chimp is being ignored by its mother, it pushes its lips out and whimpers or makes a short hooting sound until she notices it.

Pouting face

22 Colors can be used to communicate. A healthy male mandrill baboon has a brightly colored face and rump. This probably helps a female pick the best male with which to mate.

23 **Apes can use beating sounds to show they are angry.** Large male gorillas, for example, slap their cupped hands loudly against their chests. They also beat the ground with their fists, making noises that scare off other males or enemies.

Apes and monkeys use noises such as screeches and grumbles to communicate. The loud calls of these howler monkeys, which are more like roars, can echo through the forest and warn others to keep out of their territory.

24 **Howler monkeys are the loudest of all primates.** When they get together in a group, or troop, and start calling, the noise can be heard up to 2 miles away! They use this type of communication to tell other howler monkeys to keep their distance.

On the move

25 It's easy to fall when you are leaping between tall trees—and that's why many monkeys have tails that can grip branches. Some monkeys' tails work like an extra arm or leg, and can be moved and controlled in the same way we can move our four limbs.

26 New World monkeys are very similar to Old World monkeys, but they spend almost all of their lives in jungles or rain forests. Their tails are particularly useful for grabbing hold of branches and are called prehensile tails.

27 When monkeys dart around the treetops they can find plenty of food. There are fruits, nuts, seeds, insects, lizards, and birds' eggs to be found in the upper layers of a rain forest.

The tail of a wooly spider monkey has bald patches that allow for a better grip.

28 Squirrel monkeys often live in large groups with more than 50 members, but some groups may number as many as 200! These little primates have tails that are longer than their bodies and they are extremely agile and acrobatic in the treetops.

29 Squirrel monkeys are so light and nimble they can run along branches that are no more than one inch thick. They spend almost all their lives in trees and rarely come to the ground. When they do, they scurry around looking for food, or spend a little while playing.

This squirrel monkey has wrapped its tail around a tree for extra support.

WHERE IN THE WORLD?

New World monkeys come from South and Central America. Use an atlas to find which of these countries are in the Americas:

Bolivia   Brazil   Ghana   Peru
India   Sri Lanka   Panama

Answer:
Bolivia Brazil Peru Panama

Capuchin monkeys scamper through the treetops, foraging for food. They eat a wide variety of fruits. They also catch insects and other small animals.

Spider monkeys

30 If you are a monkey, it's a good idea to spend most of your time up in the trees. Hidden amongst the leaves and branches, it's easier to stay out of the reach of other animals that may want to eat you. Although big cats such as jaguars can climb trees, they can't chase monkeys onto the thinnest branches.

31 Spider monkeys are some of the fastest, most nimble of the primate climbers. Their arms, legs, and tails are extremely long, and they can hang from a tree by a single limb or tail. They usually run along a tree's branches with their tails curled over their backs, and it's rare for them to walk around on the ground.

32 Spider monkeys can travel up to 3 miles a day browsing for their favorite food of ripe fruit. During the dry season, from July to September, there is less food around, so the monkeys spend most of their day resting and saving energy.

THAT'S HANDY!

Find out just how useful your hands are!
You will need:
a big bowl of cold water • apples
Put the apples in the bowl and place it on the floor. Keep your hands behind your back and grab an apple out of the bowl, using just your mouth.
It's not easy.

High up in the forest canopy, black-handed spider monkeys hang by a hand, foot, or tail and swing at speed between branches.

33 Spider monkeys are noisy animals and make loud barking calls if they're scared. This warns the rest of the troop to beware, because a predator, such as a wild cat or snake, may be around. If some members of the family are separated, they make whinnying noises until they find one another.

34 While many primates use their thumbs to grip, some spider monkeys don't even have any! Despite being thumbless, these agile animals can hold onto branches by using a hand and four long fingers like a hook.

Baboons

35 Not all monkeys live in trees—some prefer to spend most of their time on the ground. Baboons, for example, are Old World monkeys that usually only climb trees to sleep. In fact, most baboons live in dry, rocky, or grassy places where there are few trees. Some baboons even sleep on cliff faces.

36 Baboons may not be great climbers, but they are excellent swimmers. Those that live by the sea often wade into the water to find crabs and other shelled creatures to eat.

Baboons come to waterholes and rivers to drink. They are vulnerable to attack and keep a lookout for predators between sips.

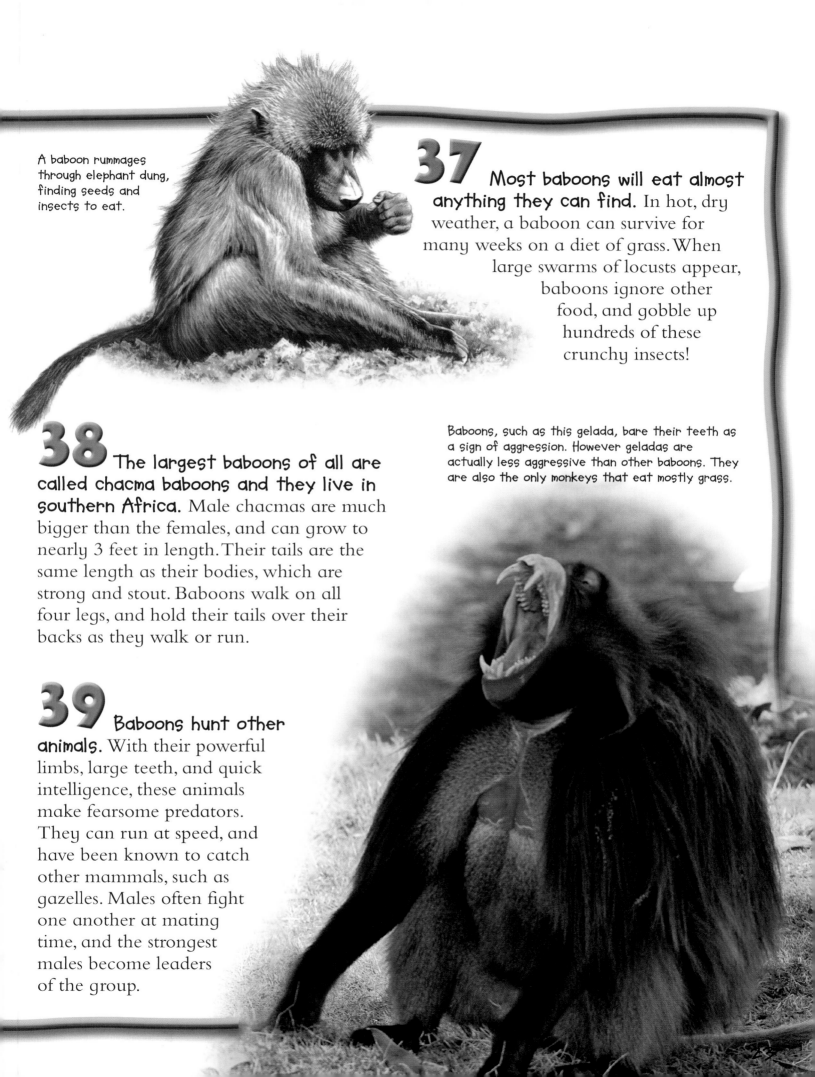

A baboon rummages through elephant dung, finding seeds and insects to eat.

37 **Most baboons will eat almost anything they can find.** In hot, dry weather, a baboon can survive for many weeks on a diet of grass. When large swarms of locusts appear, baboons ignore other food, and gobble up hundreds of these crunchy insects!

38 **The largest baboons of all are called chacma baboons and they live in southern Africa.** Male chacmas are much bigger than the females, and can grow to nearly 3 feet in length. Their tails are the same length as their bodies, which are strong and stout. Baboons walk on all four legs, and hold their tails over their backs as they walk or run.

Baboons, such as this gelada, bare their teeth as a sign of aggression. However geladas are actually less aggressive than other baboons. They are also the only monkeys that eat mostly grass.

39 **Baboons hunt other animals.** With their powerful limbs, large teeth, and quick intelligence, these animals make fearsome predators. They can run at speed, and have been known to catch other mammals, such as gazelles. Males often fight one another at mating time, and the strongest males become leaders of the group.

Macaques

40 There is one group of monkeys that has been able to make its home in all sorts of places—the macaques. There are 15 different types of macaque and they live in North Africa and Asia in forests and swamps. In these places it can be snowy, or hot and dry.

Rhesus monkeys can find food in towns, and are sometimes considered pests. This monkey is living in a temple in Thailand.

41 In the Indian state of Uttar Pradesh there are groups of macaques called rhesus monkeys. They find shelter in buildings or beneath cars in cities and towns. Like most macaques they are able to eat whatever food they come across, such as fruit, berries, insects, seeds, and flowers.

42 Crab-eating macaques live in Asian forests where the weather is always hot and rainy. At night they huddle together on branches that hang over rivers. If they get scared, they simply drop into the water and swim away!

Crab-eating macaques hunt for crabs, shrimps, frogs, and octopuses to eat in mangrove swamps.

44 **Like most other primates, macaques eat soil.** No one knows for sure why they do this but there are probably two reasons. Firstly, soil contains important minerals, such as iron, calcium, and sodium that are good for the monkeys' health. The second reason is that some soils appear to contain substances that kill small worms living inside the monkeys' bodies.

43 **The fluffy pale fur and pink face of a Japanese macaque is a startling sight.** These monkeys don't just look peculiar—they behave rather strangely too! Japanese macaques live in the cold mountains of Honshu, in Japan, but one troop has found a great way to keep warm—they bathe in the hot-water springs that bubble up from the ground!

Bathing in hot-water springs is a regular winter pastime for Japanese macaques. Having hot baths is one way to survive the extreme cold.

Fascinating faces

45 Many monkeys and apes look so like humans that we find their faces fascinating. Some primates, though, are simply odd-looking!

46 Black-and-white colobus monkeys live in Africa where they've been hunted for their amazing coats for hundreds of years. When these monkeys leap between trees, they use their long, fluffy tails to steer or change direction.

47 The proboscis monkey is a very peculiar primate. "Proboscis" means "nose" and it's easy to see how this animal got its name. Although the baby monkeys are born with normal-sized noses, they soon start to grow, and by the time they are adult males, proboscis monkeys have long, droopy snouts.

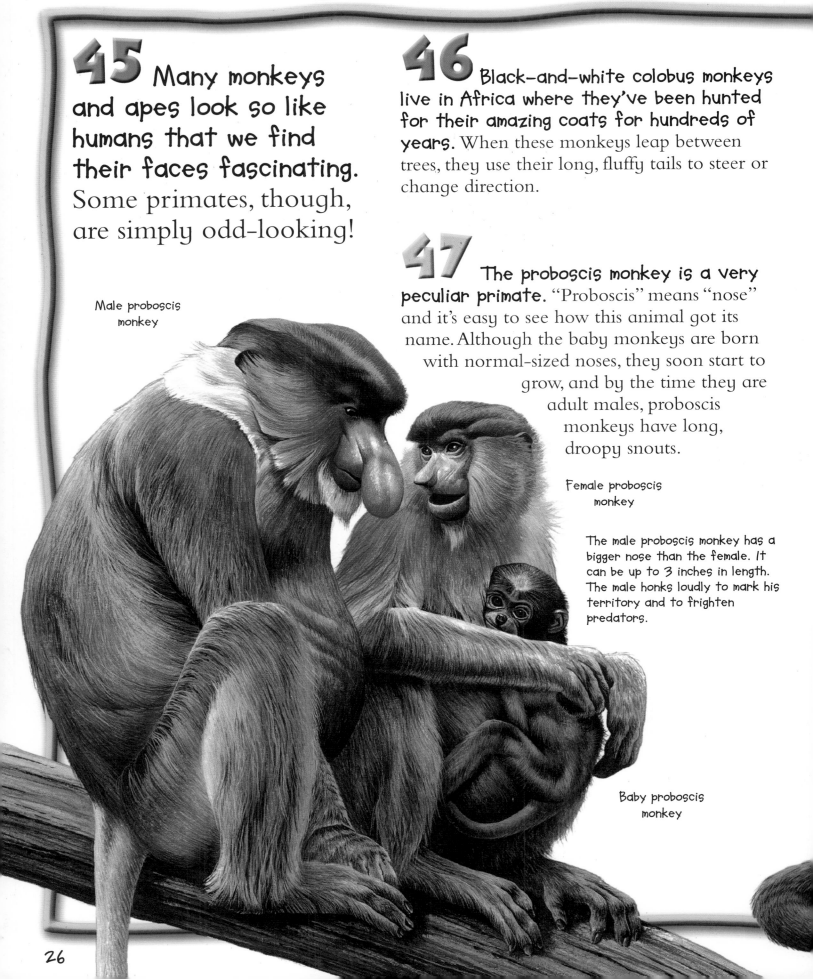

Male proboscis monkey

Female proboscis monkey

The male proboscis monkey has a bigger nose than the female. It can be up to 3 inches in length. The male honks loudly to mark his territory and to frighten predators.

Baby proboscis monkey

QUIZ

Unscramble the letters to find monkey names:

OBANOB ROGNATANU
ORILGAL PIMCH

Answers:
Baboon Orangutan
Gorilla Chimp

Red uakaris are shy and unaggressive monkeys, despite their fierce red faces.

48 **Gelada baboons may not be pretty, but they are difficult to ignore!** They have long, expressive faces and bare chests, which have bright red or pink triangular patches. These monkeys have thick brown fur on their bodies, which grows long around the shoulders, head, and neck on males.

49 **The red uakari monkey looks as if he's lost all the fur from his face, but he's meant to look this ugly!** Young red uakaris are born with normal amounts of hair on their pale pink faces, but by the time they are about two years old, their faces have become bald and red.

50 **De Brazza's monkeys look smart and tidy, with their neat white beards, dark eyebrows, and orange crowns.** These African monkeys live on a diet of fruit, flowers, seeds, leaves, and lizards. They call to one another with loud, booming sounds across the dense rain forests where they live.

De Brazza's monkeys mark their territory but they avoid intruders rather than challenge them.

27

Intelligent creatures

51 Monkeys are clever, and apes are even smarter. All primates have big brains, and they put them to good use. When faced with a problem, the most intelligent monkeys and apes may be able to work out a solution and some of them even use tools, such as sticks and stones. Chimpanzees are probably the most intelligent apes of all.

52 Many monkeys and apes use tools to help them get food. Capuchin monkeys smash tough fruits with sticks of wood until they can break into them. Youngsters often copy the adults, but they aren't usually strong enough to succeed!

Capuchin monkeys use a variety of techniques to break open nuts. This capuchin is using a rock to crack palm nuts.

HOW CLEVER ARE YOU?

Can you learn a new skill? Choose from one of these three activities and ask a grown-up to teach you how to do it:

1. Make a meal for your family.
2. Plant a garden and help it grow.
3. Learn how several words in a foreign language.

54 **Monkeys and apes are quick to learn new things.** Youngsters watch their parents and older members of the family, and copy them when they collect food or use tools. This ability to learn has been put to good use by people in Southeast Asia, who have taught pig-tailed macaques to climb palm trees and throw down coconuts for them!

Pig-tailed macaques have been trained to climb trees and collect coconuts. They scramble up the trees, twist the coconuts from the top, and drop them to the ground.

53 **Orangutans find many ways to use leaves, and have been seen using them as fly-swats, "toilet paper," and umbrellas!** These apes are so clever they can collect stacks of sticks to hold prickly fruit while they break it open. They can connect short sticks together to make one long one, make swings from rope, and they have even been seen stacking boxes on top of one another to create a ladder.

Orangutans sometimes hold leafy branches over their heads to shelter from the rain and the sun.

55 **Baboons that live in Saudi Arabia filter their own drinking water!** These clever primates dig holes in the ground next to ponds. They watch the water filter through the sandy soil before drinking it. The sand stops dirt and bugs from passing through and this means the baboons are less likely to become ill.

Family life

56 **Most monkeys and apes live in families and eat, sleep, and play together.** Primates are described as social animals. This means they prefer to stay together as groups—and they can be sociable because they are so good at communicating with one another. A group may have just two members, or several hundred.

57 **New families begin with mating.** This is when the males and females get together to start families. Male monkeys and apes often have to impress the females to persuade them to mate, and they sometimes do this by fighting off other males. It's usually the strongest and biggest males who get the most mates.

Young baboons start to explore away from their mothers when they are two or three months old.

58 **The time it takes for a baby to grow inside its mother's body is called a pregnancy.** Female monkeys and apes are pregnant for around five to nine months—animals with longer pregnancies tend to be more intelligent, but give birth to young that are quite helpless. Just like human babies, monkey infants rely on their mothers for everything.

This baby silvered langur monkey depends on its mother for everything. Its bright orange coat will last until the infant is about three months old.

59 Young primates spend a lot of time with adults, watching and learning. Youngsters watch their mothers eating plants, and learn which ones are good to eat, and which ones may be harmful. They also enjoy playing—but if a young gorilla jumps on its dad one too many times, it may get a gentle push to warn it to stop!

60 Silvered langurs have silver-gray fur, but their babies are born bright orange. After three months their striking color fades as gray hair grows. No one knows why the youngsters look so different, but it may remind older langurs to treat them more gently.

Gibbons

1. Preparing to move, a lar gibbon swings its body forward, gripping the branch with both hands.

2. One hand lets go of the branch behind as the gibbon moves forward.

61 Gibbons are the fastest-moving primates, and can swing through the trees at great speed. The swinging movement is called brachiation, and it allows gibbons to reach speeds of up to 35 miles per hour. This makes it practically impossible for any predator to catch a gibbon.

Gibbons live in hot and humid rain forests in Southeast Asia. They make death-defying leaps between trees, covering up to 50 feet at a time, at great speed.

62 Silvery gibbons may look like monkeys, but they are actually apes. Apes are more similar to humans than monkeys—not just in appearance but in intelligence and behavior too. Unlike monkeys, apes often stand or sit upright, they don't have tails, and their faces are flat. Gibbons are called "lesser apes" while their cousins—chimps, bonobos, gorillas, and orangutans—are called "great apes."

3. The gibbon prepares to grab the next branch with its free hand.

4. Without stopping for a second, the gibbon swings forward again.

63 Gibbons spend all their lives in trees and have long arms and strong shoulders to support their weight as they swing through the trees. They hold onto one branch with one hand, then swing forward as the other hand reaches out to grab the next branch. Once they get their speed up, gibbons can let go between handholds and almost fly through the trees.

64 Gibbons are light and agile. This allows them to climb to the thinnest branches and get food that other animals can't reach. Gibbons make loud calls to one another and their voices can carry for distances of one mile or more.

I DON'T BELIEVE IT!

Gibbons' feet have leathery soles and big toes that can grasp onto branches like thumbs. Even so, gibbons sometimes lose their footing and many fall, suffering painful broken bones.

Bonobos

Bonobos stroke one another and clean each other's fur. This is called grooming and it's an important part of living as a group.

65 Bonobos are secretive apes and very little is known about them. For a long time, it was thought that they were a type of chimpanzee—but now they are recognized as a different animal altogether. They have smaller heads than chimps and long, lean limbs.

66 These black-faced apes are experts at talking to one another. They use their voices and their faces to tell each other exactly how they are feeling. The noises they make include hooting, barking, and grunting. Bonobos even make a panting, laugh-like sound when they are playing or being tickled!

67 Baby bonobos depend on their mothers and stay close by them for the first few years of life. A newborn baby won't leave its mother's side for the first three months, and even when it is a year old, a young bonobo doesn't stray more than a few feet from its parent.

Baby bonobos don't eat solid food until they are about one year old. Until then, they drink their mother's milk.

68 Being part of a family is very important to a bonobo. Young males always stay close to their mothers and never leave their group. Young females, however, eventually leave when they are about seven years old and join another group. Unusually, females lead the groups, not males.

ANIMAL GROUPS

Bonobos live in groups that are led by females. Use the Internet or a library to find out more about animals that live in groups. Investigate elephants, bees, penguins, and lions.

69 Bonobos living in San Diego Zoo in America have been watched by scientists who want to find out more about how they behave. The scientists discovered that the bonobos invented their own game of "blind man's bluff." In this game, a young ape covers his eyes with his arm then tries to make his way through the play area without losing his balance or bumping into things. The others youngsters watch first, then join in the game!

Like chimps, bonobos lean on their knuckles, rather than the palms of their hands, when they walk or stand.

Clever chimps

70 **Chimpanzees are our closest living relatives, yet they are in danger of extinction.** It seems strange that humans don't do everything in their power to save the lives of their animal cousins, but chimps—the best known of all the great apes—are fighting a tough battle for survival.

71 **Chimpanzees live in the rain forests of western and central Africa.** They spend much of their time in trees, and they can swing from branches like gibbons, but not so well. When they are on the ground, chimps walk on all four limbs, and even run in this position. They can also stand up on their legs and can walk for up to one mile. Walking upright leaves their hands free for throwing stones at enemies, which they sometimes do!

Not many animals are clever enough to use tools. This chimp is using a leaf to collect water.

72 Like most monkeys and apes, chimps eat plants and insects. They visit trees laden with fruit, when it's in season, but otherwise they eat flowers, seeds, nuts, eggs, and honey. Unlike most other primates, chimps hunt other animals to eat. They also eat termites—small ant-like insects.

Chimps are 25 to 35 inches tall and weigh up to 130 pounds. They spend a lot of time in trees, looking for food.

73 Termites live in large nests, so they can be difficult to reach. Chimps have worked out a way to get to the termites by using sticks. They poke the sticks into the termite nest and, once the bugs have swarmed all over the sticks, they pull them out and eat the termites.

Chimps insert sticks into termite mounds to catch and eat the termites. Scientists have discovered that it is much harder than it looks!

74 Chimps don't just use sticks to catch termites, they use them to pull down fruit from trees. They've also learned how to use leaves to wipe down their bodies or to scoop up water to drink. Young chimps don't know how to do these things naturally —they only find out by watching adults and copying them.

QUIZ

Newborn apes are called "babies" or "young." Can you match these animals to their young, which are given special names?

goat calf elephant caterpillar butterfly foal horse kid

Answers:
goat—kid elephant—calf
butterfly—caterpillar horse—foal

37

75 Chimps may look cute, but they can be very violent animals. They live in large groups and they defend their home area, or territory, from other groups. Young male chimps even patrol the edges of their territory, looking for intruders.

An angry chimp is a terrifying sight. Male chimpanzees are amongst the most dangerous animals in the African rain forest. An adult male defending his territory may attack and kill chimps from other communities.

76 When male chimps come across strangers from other troops, they may attack them, even females. However chimps don't just attack other troops to defend their own territory—they will also invade other chimps' territories and kill them. In this way, a strong troop can take over the territory of a weaker one.

77 Chimps kill for food. They have been known to attack baboons, pigs, and hoofed animals, such as antelope. Although a chimp may have some success when it hunts on its own, it's much better to be part of a hunting group. Adult chimps have been known to kill and eat young chimps—and there have even been some reports of chimps carrying off human babies.

78 Working together is easier if you can communicate with one another, and chimps are experts at communication. If one chimp finds some food it may hoot, scream, and slap fallen tree trunks with its hands to get attention and bring others to the food. When chimps shriek or roar, the sound can travel up to one mile away!

79 Chimps don't just use noises to communicate, they also use facial expressions, kisses, and panting. Chimps in captivity have been taught how to use sign language to communicate with humans. Sign language is a method of communication that mostly uses hands, rather than sounds, to show words and ideas.

80 When adult chimps from the same troop see one another they sometimes hug, and settle down to groom each other. Chimps use their fingers to run through each other's hair and pick out lice, dirt, and twigs.

Orangutans

81 The red apes of Sumatra and Borneo are called orangutans and they are the only great apes that live in trees. These mighty animals are gentle creatures that spend most of their time alone, searching for food.

82 There are two different types of orangutan. One lives on the Indonesian island of Borneo, and the other lives on the neighboring island of Sumatra. Both types are in danger of extinction, but Sumatran orangs are not expected to survive the next 5 to 10 years.

83 Male orangutans live alone. Some have their own territory, which they protect from other males. They have large cheek pads, big heads, and masses of thick, long hair. At mating time they make loud calls that can be heard over long distances, telling the females to come and find them!

WHAT'S THE DIFFERENCE?

Male and female orangutans look very different. Find out what the males and females of these animals look like:

Peacock   Fallow deer
Mallard duck

84 Despite their great size, orangutans can climb trees, using all four limbs to grip onto branches. Once in the trees, they pick fruit, which makes up most of their diet. Although adults prefer to live alone, they will gather in groups around a fruit tree that is laden with ripe fruit.

Young orangutans stay with their mothers until they are about eight years old—longer than all other primates, except humans.

85 Sumatran orangutans share their rain forests with tigers, clouded leopards, and crocodiles. Small females and young orangutans are more likely to be attacked by these deadly hunters than adult males are.

Gorillas

86 There are three types of gorilla—western, eastern, and mountain—but they all look alike. These apes are the largest of all primates, and a mature adult male—called a silverback—can reach nearly 7 feet in height and weigh 440 pounds.

87 Although there are three different types of gorilla, they all lead very similar lives. Gorillas live in forests—either tropical rain forests or rain forests in mountainous areas that can become cold. Mountain gorillas have longer, thicker fur to keep them warm. Gorillas eat plants and spend much of their day chewing leaves, shoots, and stems.

88
At night, gorillas make themselves nests to sleep in. It only takes an adult about five minutes to break and bend twigs to create a good sleeping place. They never sleep in the same nest twice.

Gorillas like to rest after every meal and at night. They make nests to sleep in up in trees. Large males make nests on the ground.

89
Gorillas are more likely to walk away from trouble than fight. If a family can't escape an intruder, the oldest male stares at the stranger, or barks. If this doesn't work, he stands up tall, hoots, and beats his chest. He only runs at the intruder as a last resort—but he may use his huge fangs and considerable strength to kill his enemy.

Gorillas live in groups. The silverback is the leader and protector of the group. He makes all of the day-to-day decisions.

QUIZ
1. What is the largest primate?
2. What is the smallest primate?
3. What is the name of the island where lemurs live?

Answers:
1. Gorilla 2. Mouse lemur 3. Madagascar

90
About two out of every five gorillas die before they reach their first birthday. Some are killed by silverbacks from other families, but scientists don't know the reason for most of the deaths. It's most likely that the babies simply become ill.

43

Primates at risk

91 Many monkeys, and all apes, are at risk of extinction. This means that once they die out, they'll be gone forever. There are different reasons for this, but the most important one is the loss of habitat. Humans are cutting down forests, woodlands, and grasslands where primates live.

93 Orangutans live in forests that are being cut down and turned into farmland to grow palm trees. The palm trees produce oil, which can be used in foods, to make soap, and as a fuel for cars. Some of the forests are cut down for the wood that comes from the trees. Orangutans are also hunted for meat by local workers.

Large parts of the Earth's natural woodlands are being cut down. This is called "deforestation." The loss of their homes is the greatest threat to primates.

92 In Africa, gorillas and chimps are also dying out because their forests are being cut down. They have also suffered the effects of hunting over the last 200 years. At one time it was fashionable for Europeans to go to Africa, with the aim of killing as many animals as they could. Now poachers catch and kill gorillas and chimps to sell their body parts as souvenirs or for food.

94 For many years, monkeys have been taken out of the wild so that they could be used for experiments. Primates are similar to humans in many ways, so scientists test medicines on them before trying them out on people. Today, many primates are protected by laws that prevent them being taken from the wild.

A caged baby monkey in Thailand, which will probably be sold as a pet. Many primates are captured for experiments, or sold as pets.

95 There are around 700 mountain gorillas in the wild, only a few thousand Sumatran orangutans, and perhaps as little as 400 golden bamboo lemurs. The risk of extinction for most monkeys and apes is getting greater every year—and the outlook for them is bleak.

Making animals perform for human entertainment is cruel and unnecessary. In parts of Asia, monkeys are trained to perform tricks to earn money for their owners.

I DON'T BELIEVE IT!

Every year, thousands of people all over the world dress up in gorilla costumes and run 4 miles to raise money to save the last wild mountain gorillas.

A better future?

96 Although many primates are in danger of extinction, there is a chance that they could be saved. Some governments realize that looking after all of their wildlife, not just their primates, is very important. They have set aside areas called national parks, or wildlife sanctuaries, where wildlife is protected.

NATURE EXPLORER

Find out about your own local environment and wildlife. Discover the names of living things in your area, and draw pictures of them. Try to find:

Five birds
Three mammals
Four trees
Two flowers

Jane Goodall is a scientist who has spent many years studying the chimpanzees in Gombe Stream National Park, Tanzania, in Africa. In 1977 she established the Jane Goodall Institute, which is a leading organization in its efforts to protect chimpanzees and their habitats.

97 Keeping primates in zoos is not ideal, but here they can be protected and studied. Scientists who work in zoos encourage primates to breed by looking after them well. They hope to return some to the wild once their habitats have been made safe.

98 Saving primates is often in the hands of the people who live near them. Sometimes people don't realize how precious their wildlife is, or they are too poor to care for their environment. Conservationists teach local people new skills and help them to preserve their wildlife for future generations.

In Borneo, wildlife sanctuaries have been set up to try and save the remaining orangutans.

99 **Tourists pay money to come and see monkeys and apes in their natural homes.** This money helps the local people to live, and helps others from all over the world learn more about these beautiful creatures.

100 **Monkeys and apes are worth saving.** They are our closest relatives, yet we've brought them close to the point of extinction. It is our responsibility to undo the damage we've already done and allow the populations of these intelligent animals to thrive.

Index